FM 100-14

Field Manual
No. 100- 14

Headquarters
Department of the Army
Washington, DC, 23 April 1998

Contents

	Page
Preface	ii
Introduction	iii
Chapter 1 Risk Management Fundamentals	1-1
Background	1-2
Principles	1-3
Applicability	1-4
Constraints	1-7
Chapter 2 Risk Management Process	2-0
The Five Steps: An Overview	2-0
The Five Steps Applied	2-2
Tools and Pitfalls	2-19
Chapter 3 Risk Management Implementation	3-0
Moral and Ethical Implications for Leaders	3-0
Responsibilities	3-1
Integration into Training and Operations	3-7
Assessment of the Risk Management Process	3-9
Appendix Examples of Risk Management Application	Appendix-1
Glossary	Glossary-0
References	References-0
Index	Index-1

DISTRIBUTION RESTRICTION: Approved for public release; distribution is unlimited.

Preface

FM 100-14 applies across the wide range of Army operations. It explains the principles, procedures, and responsibilities to successfully apply the *risk management process* to conserve combat power and resources. The manual applies to both Army and civilian personnel during all Army activities, including joint, multinational, and interagency environments.

The manual is intended to help commanders,[1] their staffs, leaders,[2] and managers develop a framework to make risk management a routine part of planning, preparing, and executing operational missions and everyday tasks. This framework will allow soldiers to operate with maximum initiative, flexibility, and adaptability. Although the manual's prime focus is the operational Army, the principles of risk management apply to all Army activities.

Army operations—especially combat operations—are demanding and complex. They are inherently dangerous, including tough, realistic training. Managing risks related to such operations requires educated judgment and professional competence. The risk management process allows individuals to make informed, conscious decisions to accept risks at acceptable levels.

This manual is not a substitute for thought. Simply reading it will not make one adept in building protection around a mission.[3] Soldiers should compare the doctrine herein against their own experience and think about why, when, and how it applies to their situation and area of responsibility. If the doctrine herein is to be useful, it must become second nature.

The proponent of this manual is HQ TRADOC. Send comments and recommendations on DA Form 2028 directly to Commander, US Army Training and Doctrine Command, ATTN: ATBO-SO, Fort Monroe, VA 23651-5000.

Unless this publication states otherwise, masculine nouns and pronouns do not refer exclusively to men.

---

1 The term *commander* as used herein refers to personnel in a command position.

2 The term *leader* as used herein refers to commanders, personnel in the chain of command (team, squad, section, platoon leader), and staff members having personnel supervisory responsibility.

3 The term *mission* as used herein includes mission, operation, or task.

Introduction

Risk management is not an add-on feature to the decision-making process but rather a fully integrated element of planning and executing operations... Risk management helps us preserve combat power and retain the flexibility for bold and decisive action. Proper risk management is a combat multiplier that we can ill afford to squander.

General Dennis J. Reimer
Chief of Staff, Army
27 July 1995

The Army's fundamental purpose is to fight and win the nation's wars. For this purpose, the country gives the Army critical resources, including those most valuable—its sons and daughters. The Army uses its resources to generate overwhelming combat power to fight and win quickly, decisively, and with minimal losses. The Army's inherent responsibility to the nation is to protect and preserve its resources—a responsibility that resides at all levels. Risk management is an effective process for preserving resources. It is not an event. It is both an art and a science. Soldiers use it to identify tactical and accident risks, which they reduce by avoiding, controlling, or eliminating hazards.

The Army introduced the risk management process into training, the operational environments, and materiel acquisition in the late 1980s. Risk management was originally perceived as solely a safety officer function. However, by the early 1990s, the Army established a goal to integrate risk management into all Army processes and activities and into every individual's behavior, both on and off duty. Since the process was introduced, the personal involvement of commanders in preventing accidents—and their aggressive use of the process—have become driving factors in the steady downward trend in Army accidental losses.

Leaders must understand the importance of the process in conserving combat power and resources. Risk management, like reconnaissance and security, is an ongoing process that continues from mission to mission. Within the mission, leaders must know when the process begins and who has responsibility. It must be integral to the military decision. The process is an important means to enhance situational awareness.

Risk decisions are commanders' business. Such decisions are normally based on the next higher commander's guidance on how much risk he is willing to accept and delegate for the mission. Risk decisions should be made at the lowest possible level, except in extreme circumstances. Training operations, including those at combat training centers (CTCs), may be of such intensity that risk decision are retained at a higher level.

Both leaders and staffs manage risk. Staff members continuously look for hazards associated with their areas of expertise. They then recommend controls to reduce risks. Hazards and the resulting risks may vary as circumstances change and experience is gained. Leaders and individual soldiers become the assessors for ever-changing hazards such as those associated with environment (weather; visibility; contaminated air, water, and soil), equipment readiness, individual and unit experience, and fatigue. Leaders should advise the chain of command on risks and risk reduction measures.

Index

A

accident risks 2-2, 2-4, 2-11, 3-1
adversaries 2-7
after-action reviews 3-10
ambiguity 1-2
antiterrorism 3-10
assessment 3-9
avenues of approach 2-5
avoidance controls. *See* controls, types of
awareness. *See* situational awareness

B

battlefield controls. *See* controls

C

chain of command 3-2
civilians 2-1, 2-7
climate 2-6
COA
- analysis 2-3, 2-7
- development 2-3, 2-7, 2-8
- enemy 2-4
- evaluation 1-3

combat operations ii
combat power ii, iii, 1-1, 2-2, 2-6, 2-12, 2-13
- conserving 2-0
- critical 2-13
- enemy deployment of 2-5
- friendly 2-5
- loss of 2-9

command climate 3-2
commander
- and identification of METL 2-19
- and the risk management process 2-19
- responsibilities of iv, 2-3, 2-16, 3-2–3-5

commander's guidance 2-16, 3-2, 3-4
complacency 3-3
constraints 1-7
contract support personnel 2-1
control measures. *See* controls
controls iv, 1-3, 1-7, 2-3, 2-7
- acceptability of 2-14
- alternate 2-16
- approval of 1-3
- battlefield 2-15
- conversion of 2-17
- coordination and communication 2-17
- criteria for (Fig 2-6) 2-14
- development of 2-0, 2-13, 2-14, 3-4
- effectiveness of 2-2, 2-19
- emplacement of 2-18
- enforcement of 2-18
- examples of 2-15–2-16
- execution of 2-17, 3-1, 3-5
- feasibility of 2-14
- implementation of 2-0, 2-14, 2-17, 2-18
- ineffective 2-19, 2-21
- integration into OPORDs 1-3
- modification of 2-18
- monitoring of 2-18

reassessment of 2-16
and residual risk 2-16
and risk decision 2-16
safety and environmental 1-6
suitabiity of 2-14
and tools for development 2-19
types of 2-14
counterinsurgency operations 2-6
counterintelligence 2-4. *See also* intelligence
courses of action. *See* COA
cover and concealment 2-5

D

danger, sources of 1-2, 2-3, 3-3
decision levels 1-3
decisive terrain 2-5. *See also* terrain
degree of severity. *See* severity, degree of
discipline 2-18

E

educational controls. *See* types of controls
enemy, as part of METT-T 2-3. *See also* combat power
environment
damage to 1-2
impact on 2-7
evaluation 3-9
evaluation, of risk management process 2-18
examples of risk management work sheets
for battalion/task force (Fig A-5) Appendix-5–7
for company/team (Fig A-4) Appendix-4
for division/corps/EAC (Fig A-6) Appendix-8–10
for squad/platoon (Fig A-3) Appendix-3
exposure to hazards 2-2, 2-7, 2-15

F

fatigue 1-2, 2-6
fear 1-2, 3-3
foreign nationals 2-7
fragmentary order. *See* FRAGO
FRAGO 2-3
fratricide 1-2, 2-15, 3-4
friction 1-2, 3-3
friendly fire. *See* fratricide
friendly vulnerabilities 2-4

G

geographical area, acclimatization to 2-6

H

hazard, definition of 2-2
hazards iii, iv, 3-1, 3-6
and accident risks 2-2
assessment of 2-7, 3-10
and common shortfalls 2-4
control of 3-4
heat and cold 2-5
identifying and controlling 2-0, 2-2, 2-3, 3-6, 3-9
long-term 2-18
outside of assigned responsibilities 3-4
review of 3-10
and tactical risk 2-2
variable 2-18, 3-1
health considerations 2-6
hostile elements. *See* adversaries

I

implementation of controls 2-17
incident, credibility of 2-7
individuals, responsibilities of 3-7
inexperience 2-6
infrastructures, collecting information on 2-4
instructions for work sheet Appendix-1
insurgency operations 2-6
integration of risk management into MTP task (Fig A-7) Appendix-11–17
intelligence. *See also* counterintelligence
  collection plan Appendix-15
  dissemination of 2-4
  estimates Appendix-12
  and identification of hazards 2-4
intelligence-preparation-of-the-battlefield. *See* IPB
IPB 2-4

K

key terrain 2-5. *See also* terrain
knowledge, lack of 2-6

L

leaders. *See also* supervisors
  and assessment of hazards 2-8
  and avoidance 2-14
  and continuous planning 2-18
  and development of controls 2-13
  and discipline 2-18
  and estimated degree of severity 2-11
  and evaluation of risk management process 2-18, 2-19
  and implementation of controls 2-17
  and METT-T 2-3
  and mission analysis 2-3
  and mission preparation 2-17
  and mission rehearsal and execution 2-18
  and modification of controls 2-18
  and moral and ethical implications 3-0
  and overlap tours 2-16
  and overwatch of controls 2-18
  responsibilities of iii, 2-1, 2-2, 2-3, 3-5
  and risk assessment 2-16
  and situational awareness 2-18
  and time available 2-6
  and tools to assess identified hazards 2-19
  and troops 2-6
  and weather 2-5
  as supervisors 2-17
  unit 2-3
leadership 2-6, 2-14, 3-2
lessons learned 2-10, 3-10
levels of risk 2-11, (Fig 2-5) 2-12, 2-13
  extremely high 2-12
  high 2-12
  low 2-13
  moderate 2-12
LOGCAP 2-17, 3-8
Logistics Civil Augmentation Program. *See* LOGCAP
losses 1-2. *See also* fratricide

M

manning levels 2-6

matrix. *See* risk assessment matrix

media, hazards to 2-7

METL 3-9

- and degree of related risk 2-19
- identification of 2-19
- -related missions 2-6

METT-T 1-6, 2-0, 2-3, 2-12, 2-21, 3-2

military decision-making process 2-0(Fig 2-1), 2-1, 2-3, 2-7, 2-13

minimizing risk 3-1

mission

- analysis 2-3, 2-7
- failure, causes of 2-6
- as part of METT-T 2-3
- receipt of 2-3

mission, enemy, terrain, troops, and time available. *See* METT-T

mission training plan. *See* MTP

missions, long-term 2-6

morale 2-6

moral and ethical implications 3-0–3-1

MTP

- integrating risk management into Appendix-11
- and risk management considerations 2-19

multidiscipline counterintelligence analysis 2-4

N

nation building 2-6

NGOs 2-7, 2-17, 3-8

nongovernmental organizations. *See* NGOs

O

observation and fields of fire, cover and concealment, obstacles, key terrain and decisive terrain, and avenues of approach. *See* OCOKA

OCOKA 2-5

one-third/two-thirds rule 2-6

operating environment 1-2, 1-3

operational deployment 3-1

operational hazards 1-2

operational tempo. *See* OPTEMPO

OPLANs 1-3, 3-10

OPORDs 3-10

OPTEMPO 1-3, 3-3

overall risk 2-7, 2-10

overlap tours 2-16

overwatch of controls 2-18

P

peacekeeping 2-6

peacetime environment 2-7

perception, and leaders' decisions 1-1

personnel tempo. *See* PERSTEMPO

personnel turnover 2-6

PERSTEMPO 1-3

physical controls. *See* types of controls

pitfalls of risk management 2-19

planning time 2-6

private voluntary organizations. *See* PVOs

probability 1-1, 2-7, 2-8, 2-10, 2-11, 2-12, 2-13, 2-15, 2-16, 2-19, 3-10

- degrees of 2-8
- frequent 2-8, 2-12

of hazards (Fig 2-2) 2-8, 2-9
levels 2-8
likely 2-12, 2-13
occasional 2-12
seldom 2-12, 2-13
unlikely 2-12, 2-13
proficiency 1-1
PVOs 2-7, 2-17, 3-8

R

readiness 3-10
reconnaissance activities 2-4
residual risk 1-3, 2-14, 2-16, 2-19, 2-21, 3-6
definition of 2-16
tactical 3-4
resources ii, iii, 1-1
resources, conservation of iii
responsibilities of commanders, leaders, staffs, and individuals 3-1–3-7
risk
definition of 1-1, 2-7
estimate of 2-7
levels of 2-10
perception of 1-1
risk assessment 2-2, 2-4, 2-7, 2-11, 2-19, 2-21, 3-6
card 2-21, 3-2
matrix (Fig 2-4) 2-11, 2-16
process 3-8
tools 3-2
risk controls. *See* controls
risk decision iv, 1-3, 2-0, 2-1, 2-13, 2-16, 2-19, 2-21, 3-2, 3-3, 3-4, 3-5
risk denial syndrome 3-2
risk levels. *See* risk, levels of
risk management 1-1, 1-7, 2-16, 3-4, 3-5, 3-6
applicability 1-4 – 1-6
background of 1-1–1-2
and the chain of command 3-0
continuous application of (Fig 2-7) 2-20
examples of application, Appendix-1–Appendix-12
fundamentals 1-2–1-7
implementation of 3-0–3-10
integrating into planning, preparation, and execution iii, 1-3
key aspects of 1-4
principles of 1-3
and regulatory and legal requirements 1-7
risk management process ii, 1-1, 1-3, 2-0–2-21, 3-1, 3-2, 3-5
assessment of 3-9–3-10
and constraints 1-7
continuous application of 2-19
the five steps 2-0
introduction of iii
risk management steps 2-1 (Fig 2-1), 2-2
risk management work sheet instructions (Fig A-1) Appendix-1
risk reduction controls 3-6

S

safety programs 3-10
sample risk management work sheet (Fig A-2) Appendix-2
severity 1-1, 2-7, 2-9, 2-10, 2-11, 2-15, 2-16, 2-19, 3-10
catastrophic 2-9, 2-10
critical 2-10
degree of 2-9
of hazards (Fig 2-3) 2-10

marginal 2-10
negligible 2-10
situational awareness iii, 2-2, 2-18, 2-19, 3-3, 3-7, 3-9
soldiers, health of 2-6
soldiers' responsibilities. *See* individuals, responsibilities of
SOPs 3-10
stability operations 2-6, 2-18
staffs, responsibiity of 2-11, 3-6
steps, of the risk management process
Step 1. Identify Hazards 2-2–2-7
Step 2. Assess Hazards 2-7–2-13
Step 3. Develop Controls and Make Risk Decisions 2-13–2-16
Step 4. Implement Controls 2-17
Step 5. Supervise and Evaluate 2-17–2-19
supervise and evaluate. *See* steps, of the risk management process
supervision techniques 2-18
supervisors, and examples of control implementation 2-17
supplies and services 2-6
sustained operations 2-18, 3-3

T

tactical risk 2-2, 2-4, 2-11, 3-1
task organization 2-6
terrain 2-5. *See also* key terrain
terrain and weather 2-4
aspects of 2-5
as part of METT-T 2-4
threat capabilities and vulnerabilities 2-4
time available, as part of METT-T 2-6
tools
to assess identified hazards 2-19
for risk management 2-19
tracking Appendix-1
training 3-1
assessment 3-9
levels of 2-6
plan 3-10
training management cycle 3-9
training and operations, integration into 3-7–3-9
troops, care of 2-6

U

uncertainty 1-2, 2-11

V

vehicles, maintenance of 2-6
violence, and risk decisions 3-3

W

wartime environment 2-7
weather hazards, identification of 2-5
work sheet instructions (Fig A-1) Appendix-1

Chapter 1

Risk Management Fundamentals

Sizing up opponents to determine victory, assessing dangers and distances is the proper course of action for military leaders.

Sun Tzu, The Art of War, "Terrain"

Risk management is the process of identifying, assessing, and controlling risks arising from operational factors and making decisions that balance risk costs with mission benefits. Leaders and soldiers at all levels use risk management. It applies to all missions and environments across the wide range of Army operations. Risk management is fundamental in developing confident and competent leaders and units. Proficiency in applying risk management is critical to conserving combat power and resources. Commanders must firmly ground current and future leaders in the critical skills of the five-step risk management process.

Risk is characterized by both the probability and severity of a potential loss that may result from hazards due to the presence of an enemy, an adversary, or some other hazardous condition. Perception of risk varies from person to person. What is risky or dangerous to one person may not be to another. Perception influences leaders' decisions. A publicized event such as a training accident or a relatively minor incident may increase the public's perception of risk for that particular event and time—sometimes to the point of making such risks unacceptable. Failure to effectively manage the risk may make an operation too costly—politically, economically, and in terms of combat power (soldiers lives and equipment). This chapter presents the background, principles, applicability, and constraints relating to the risk management process.

BACKGROUND

Throughout the history of armed conflict, government and military leaders have tried to reckon with the effect of casualties on policy, strategy, and mission accomplishment. Government and military leaders consider battle losses from different perspectives. However, both must balance the following against the value of national objectives:

- Effects of casualties.
- Impact on civilians.
- Damage to the environment.
- Loss of equipment.
- Level of public reaction.

War is inherently complex, dynamic, and fluid. It is characterized by uncertainty, ambiguity, and friction. *Uncertainty* results from unknowns or lack of information. *Ambiguity* is the blurring or fog that makes it difficult to distinguish fact from impression about a situation and the enemy. *Friction* results from change, operational hazards, fatigue, and fears brought on by danger. These characteristics cloud the operating environment; they create risks that affect an army's ability to fight and win. In uncertainty, ambiguity, and friction, both danger and opportunity exist. Hence, a leader's ability to adapt and take risks are key traits. Chapter 2 of FM 100-5 provides information on the challenging circumstances of military operations during conflict.

Historically, the Army has had more accidental losses, including fratricide (friendly fire), than losses from enemy action. See Figure 1-1. These accidental losses are the same types experienced in peacetime

Army	World War II 1942–1945	Korea 1950–1953	Vietnam 1965–1972	Desert Shield/ Storm[1] 1990–1991
Accidents	56%	44%	54%	75%
Friendly Fire	1%	1%	1%	5%
Enemy Action	43%	55%	45%	20%

[1]These numbers include the relatively long buildup time and short period of combat action

Figure 1-1. Battle and Nonbattle Casualties

during training exercises. These losses are not caused by the enemy or an adversary. Factors include—

- An ever-changing operational environment.
- Effects of a fast-paced, high-operational tempo (OPTEMPO) and a high-personnel tempo (PERSTEMPO) on unit and human performance. Examples include leader or soldier error or failure to train or perform to standards.
- Equipment failure, support failure, and the effects of the physical environment.

PRINCIPLES

The basic principles that provide a framework for implementing the risk management process are—

- *Integrating risk management into mission planning, preparation, and execution.* Leaders and staffs continuously identify hazards and assess both accident and tactical risks. They then develop and coordinate control measures. They determine the level of residual risk for accident hazards in order to evaluate courses of action (COAs). They integrate control measures into staff estimates, operation plans (OPLANs), operation orders (OPORDs), and missions. Commanders assess the areas in which they might take tactical risks. They approve control measures that will reduce risks. Leaders ensure that all soldiers understand and properly execute risk controls. They continuously assess variable hazards and implement risk controls.
- *Making risk decisions at the appropriate level in the chain of command.* The commander should address risk guidance in his commander's guidance. He bases his risk guidance on established Army and other appropriate policies and on his higher commander's direction. He then gives guidance on how much risk he is willing to accept and delegate. Subordinates seek the higher commander's approval to accept risks that might imperil the next higher commander's intent.
- *Accepting no unnecessary risk.* Commanders compare and balance risks against mission expectations and accept risks only if the benefits outweigh the potential costs or losses. Commanders alone decide whether to accept the level of residual risk to accomplish the mission.

APPLICABILITY

Risk management applies to all situations and environments across the wide range of Army operations, activities, and processes. Risk management is useful in developing, fielding, and employing the total Army force. Figure 1-2 summarizes the key aspects of risk management.

DEVELOPMENT

Development concerns include force design, manpower allocation, training and training developments, and combat and materiel developments (equipment and weapons systems) and battle laboratories.

Risk management assists the commander or leader in—

- Conserving lives and resources and avoiding unnecessary risk.
- Making an informed decision to implement a COA.
- Identifying feasible and effective control measures where specific standards do not exist.
- Providing reasonable alternatives for mission accomplishment.

Risk management does not—

- Inhibit the commander's and leader's flexibility and initiative.
- Remove risk altogether, or support a zero defects mindset.
- Require a GO/NO-GO decision.
- Sanction or justify violating the law.
- Remove the necessity for standard drills, tactics, techniques, and procedures.

Figure 1-2. Key Aspects of Risk Management

Force Design

Concerns include risks introduced in trade-off decisions that involve the design and equipping of—

- Tables of organization and equipment (TOE).
- Modification tables of organization and equipment (MTOE).
- Tables of distribution and allowances (TDA) organizations.

Manpower Allocations

Concerns include shortfalls in manning that put unit readiness and full use of combat system capabilities at risk.

Training and Training Developments

Concerns include hazardous and critical training tasks and feasible risk reduction measures that provide leaders with the flexibility to safely conduct tough, realistic training.

Combat and Materiel Developments and Battle Laboratories

Concerns include providing a means to assist in making informed trade-off decisions such as—

- Balancing equipment form, fit, and function.
- Balancing the durability and cost of equipment and spare parts against their reliability, availability, and maintainability requirements.
- Determining the environmental impact.
- Determining whether to accept systems with less than the full capabilities prescribed in requirement documents and experimental procedures.

ARs 70-1 and 385-16 and MIL-STD-882 provide details on risk management application in the Army materiel acquisition process.

FIELDING

Fielding concerns include personnel assignments, sustainment and logistics, training, and base operations.

Personnel Assignments

Concerns include making informed decisions in assigning replacement personnel. For example, a risk is associated with assigning a multiple launch rocket system crewmember as a replacement for a tube artillery cannon crewmember.

Sustainment and Logistics

Concerns include enhancing one's ability to determine support requirements, the order in which they should be received, and the potential impact of logistics decisions on operations.

Training

Concerns include helping leaders determine the—

- Balance between training realism and unnecessary risks in training.
- Impact of training operations on the environment.
- Level of proficiency and experience of soldiers and leaders.

Base Operations

Concerns include prioritizing the execution of base operations functions to get the most benefit from available resources. Examples include allocating resources for pollution prevention, correcting safety and health hazards, and correcting violations of environmental protection regulations. FM 20-400 provides specific guidance on environmental protection in military operations.

EMPLOYMENT

Employment concerns include force protection and deployment, operations, and redeployment.

Force Protection

Concerns include developing a plan that identifies threats and their associated hazards and balancing resource restraints against the risk.

Deployment, Operations, and Redeployment

Concerns include—

- Analyzing the factors of mission, enemy, terrain, troops, and time available (METT-T) to determine both tactical and accident risks and appropriate risk reduction measures.
- Determining the correct units, equipment composition, and sequence.
- Identifying controls essential to safety and environmental protection.

CONSTRAINTS

Risk management does not convey authority to violate the law-of-land warfare or deliberately disobey local, state, national, or host nation laws. It does not justify ignoring regulatory restrictions and applicable standards. Neither does it justify bypassing risk controls required by law, such as life safety and fire protection codes, physical security, transport and disposal of hazardous material and waste, or storage of classified material. Commanders may not use risk management to alter or bypass legislative intent. However, when restrictions imposed by other agencies adversely affect the mission, planners may negotiate a satisfactory COA if the result conforms to the legislative intent.

Risk management assists the commander in complying with regulatory and legal requirements by—

- Identifying applicable legal standards that affect the mission.
- Identifying alternate COAs or alternate standards that meet the intent of the law.
- Ensuring better use of limited resources through establishing priorities to correct known hazardous conditions that will result in projects with the highest return on investment funded first.

Chapter 2

Risk Management Process

First reckon, then risk

Field Marshal Helmuth von Moltke

This chapter provides the essence of the five-step risk management process. It illustrates the application of each step to military operations through the factors of METT-T.

THE FIVE STEPS: AN OVERVIEW

Risk management is the process of identifying and controlling hazards to conserve combat power and resources. The five steps of risk management are—

- Step 1. Identify hazards.
- Step 2. Assess hazards to determine risks.
- Step 3. Develop controls and make risk decisions.
- Step 4. Implement controls.
- Step 5. Supervise and evaluate.

This five-step process is integrated into the military decision-making process as shown in Figure 2-1.

FM 100-40 provides insight into the context in which the risk management process is applied herein. Areas of particular interest in FM 100-40 include—

- Solving tactical problems (Chapter 1).
- The science and art of tactics (Chapter 1).
- Hasty versus deliberate operations (Chapter 1).
- The plan-prepare-execute cycle (Chapter 1).
- Basic tactical control measures (Chapter 2).
- The factors of METT-T (Chapter 2).

Military Decision-Making Process	Risk Management Steps				
	Step 1 Identify Hazards	**Step 2** Assess Hazards	**Step 3** Develop Controls and Make Risk Decision	**Step 4** Implement Controls	**Step 5** Supervise and Evaluate
Mission Receipt	X				
Mission Analysis	X	X			
COA Development	X	X	X		
COA Analysis	X	X	X		
COA Comparison			X		
COA Approval			X		
Orders Production				X	
Rehearsal[1]	X	X	X	X	X
Execution and[1] **Assessment**	X	X	X	X	X

[1]All boxes are marked to emphasize the continued use of the risk management process throughout the mission

Figure 2-1. Risk Management Steps Correlated with Military Decision-Making Tasks

Risk decisions should be based upon awareness rather than mechanical habit. Leaders should act on a keen appreciation for the essential factors that make each situation unique instead of from conditioned response. Throughout the entire operational continuum, the commander must consider US Government civilians and contract support personnel in his risk management process. Hazards can exist, regardless of enemy or adversary actions, in areas with no direct enemy contact and in areas outside the enemy's or adversary's influence. The two types of risk that exist across the wide range of Army operations are *tactical risks* and *accident risks.*

- *Tactical risk* is risk concerned with hazards that exist because of the presence of either the enemy or an adversary. It applies to all levels of war and across the spectrum of operations.
- *Accident risk* includes all operational risk considerations other than tactical risk. It includes risks to the friendly force. It also includes risks posed to civilians by an operation, as well as an operations impact on the environment. It can include activities associated with hazards concerning friendly personnel, civilians, equipment readiness, and environmental conditions.

STEPS 1 AND 2

Steps 1 and 2 together comprise the risk assessment. In Step 1, individuals identify the hazards that may be encountered in executing a mission. In Step 2, they determine the direct impact of each hazard on the operation. The risk assessment provides for enhanced situational awareness. This awareness builds confidence and allows soldiers and units to take timely, efficient, and effective protective measures.

STEPS 3 THROUGH 5

Steps 3 through 5 are the essential follow-through actions to effectively manage risk. In these steps, leaders balance risk against costs—political, economic, environmental, and to combat power—and take appropriate actions to eliminate unnecessary risk. During execution, as well as during planning and preparation, leaders continuously assess the risk to the overall mission and to those involved in the task. Finally, leaders and individuals evaluate the effectiveness of controls and provide lessons learned so that others may benefit from the experience.

THE FIVE STEPS APPLIED

STEP 1. IDENTIFY HAZARDS

A *hazard* is an actual or potential condition where the following can occur due to exposure to the hazard:

- Injury, illness, or death of personnel.
- Damage to or loss of equipment and property.
- Mission degradation.

Hazards are sources of danger or risks due to enemy or adversary presence and other conditions not due to enemy or adversary capabilities. Hazards are found in all operational environments. Combat operations, stability operations, base support operations, and training present unique hazards for units involved in these kinds of missions. Hazards are identified during the first four steps of the military decision-making process: *mission receipt, mission analysis, COA development*, and *COA analysis*.

The ability of unit leaders and staffs to identify hazards is key. One reality of today's missions is that the aspect of a hazard can change rapidly. Things of little risk initially can quickly become major threats due to unforeseen natural or man-made events. Leaders should be aware of this possibility. Complacency to the fact that existing controls may not continue to control hazards in rapidly changing situations should be viewed as a hazard in itself.

The factors of METT-T provide a sound framework for identifying hazards when planning, preparing, and executing operations. When applying risk management to METT-T during mission analysis, leaders and staffs should look for hazards that affect both tactical and accident risks. They must identify all hazards that may present significant risks to the mission.

Mission

Leaders first analyze the assigned mission. They look at the type of mission to be accomplished and consider possible subsequent missions. Certain kinds of operations are inherently more dangerous than others. For example, a deliberate frontal attack, because of the associated movement, is more likely to expose a unit to losses than would a defense from prepared positions. Identifying missions that routinely present great risk is imperative. Leaders also look for hazards associated with complexity of the plan such as—

- A scheme of maneuver that is difficult to understand or too complex for accurate communications down to the lowest level.
- The impact of operating under a fragmentary order (FRAGO).

Enemy

Commanders look for enemy capabilities that pose significant hazards to the operation. For example, "What can the enemy do to defeat my operation?" Common shortfalls that can create hazards during operations against an enemy include failure to—

- Assess potential advantages to the enemy provided by the battlefield environment.
- Fully assess the enemy's capabilities.
- Understand enemy collection capabilities and friendly vulnerabilities to those capabilities.
- Accurately determine the enemy's probable COAs.
- Plan and coordinate active ground and aerial reconnaissance activities.
- Disseminate intelligence about the enemy to lower levels.
- Identifying terrorist threats and capabilities.

Intelligence plays a critical part in identifying hazards associated with tactical risk. Intelligence-preparation-of-the-battlefield (IPB) is a dynamic staff process that continually integrates new information and intelligence that ultimately becomes input to the commander's risk assessment process. Intelligence assists in identifying hazards during operations by—

- Identifying the opportunities and constraints the battlefield environment offers to threat and friendly forces.
- Thoroughly portraying threat capabilities and vulnerabilities.
- Collecting information on populations, governments, and infrastructures.

FMs 34-130 and 34-60, respectively, provide detailed information on IPB and on counterintelligence operations and multidiscipline counterintelligence analysis.

Terrain and Weather

In addition to those due to the enemy or adversaries, the most obvious hazards to military operations are due to terrain and weather. Terrain and weather affect the type of hazard encountered. When the enemy uses terrain to his advantage, the risk is clearly tactical. The aspects of terrain and weather may create situations where accident risks predominate. When looking at this from a purely mission perspective, familiarity of the unit with the terrain and its associated environment must be paramount. Basic issues include—

- How long the unit has operated in the environment and climate.
- Whether the terrain has been crossed before.

Terrain. The five main military aspects of terrain—*observation and fields of fire, cover and concealment, obstacles, key terrain and decisive terrain, and avenues of approach (OCOKA)*—can be used to identify and assess hazards impacting on friendly forces. Chapter 2 of FM 100-40 has details on OCOKA. The terrain analysis includes both map and on-the-ground reconnaissance to identify how well unit capabilities and mission demands can be accommodated by the terrain.

- *Observation and fields of fire.* Hazards associated with this usually involve when the enemy will be able to engage a friendly unit and when friendly unit weapons capabilities allow it to effectively engage the enemy.
- *Cover and concealment.* Hazards associated with cover and concealment are created by the enemy's ability to place direct or indirect fire on friendly forces.
- *Obstacles.* Hazards associated with obstacles may be accident or tactical. They may be due to natural conditions such as rivers or swamps or man-made such as minefields or built-up areas.
- *Key terrain and decisive terrain.* Hazards are a marked advantage terrain provides to the enemy if he controls such terrain or denies its use to friendly forces.
- *Avenues of approach.* Hazards associated with avenues of approach can affect both tactical and accident risks. Such hazards include conditions where an avenue of approach impedes deployment of friendly combat power or where it supports deployment of enemy combat power.

Weather. Weather works hand-in-hand with terrain to create hazards. To identify weather hazards, leaders and soldiers must assess the impact on operating systems. Mistakes include not considering the—

- Adverse effects of heat and cold hazards on the performance of soldiers.
- Effects of climate and weather on maintenance of vehicles and equipment before beginning an operation.
- Hazardous effects of weather on the five military aspects of terrain.

Troops

Leaders analyze the capabilities of available friendly troops. Associated hazards impact both the soldier and unit. Key considerations are level of training, manning levels, the condition and maintenance of vehicles and equipment, morale, availability of supplies and services, and the physical and emotional health of soldiers. Leaders and soldiers must be vigilant to the fact that hazards in these areas can adversely affect a mission, even when all tactical considerations point to success. Mission failure can be caused by—

- *Hazards to the physical and emotional health of soldiers.* Inadequate sanitation facilities, water purification capabilities, medical attention, and evacuation capabilities are key hazards that can arise from incomplete logistical planning. Care of troops requires long-range projections of all classes of supply, with close monitoring of mission changes that could impact availability or depletion of supplies. When beginning an operation immediately upon arriving in theater, hazards include not implementing measures to help soldiers overcome fatigue or acclimatize them to the geographical area and associated climate.
- *Hazards to task organization or units participating in an operation.* Hazards include how long units have worked together under a particular command relationship. During stability operations, task organizations may change often. Hazards include poor communication, unfamiliarity with higher headquarters SOPs, and insufficient combat power to accomplish the mission.
- *Hazards associated with long-term missions.* Long-term missions include nation building, peacekeeping, or insurgency/counterinsurgency operations. Hazards associated with these missions include the turmoil of personnel turnover, lack of continuity of leadership, inexperience, and lack of knowledge of the situation and the unit's operating procedures. An especially insidious hazard is critical-skills atrophy that results from not performing METL-related missions.

Time Available

The hazard is insufficient time to plan, prepare, and execute operations. Planning time is always at a premium. Leaders routinely apply the one-third/two-thirds rule to ensure their subordinate units are given maximum time to plan. Failure to accomplish a mission on time can result in shortages of time for subordinate and adjacent units to accomplish their mission.

Civilians

The commander's legal responsibility is to consider hazards to, and safeguarding of, civilians in his area of operations. *Civilians* include nongovernmental organizations (NGOs), private voluntary organizations (PVOs), US Government civilians, foreign national civilians, the media, and dislocated civilians put at risk by military operations. The commander must consider hazards that can occur across the range of operations, such as—

- *In a wartime environment.* The commander must consider the hazard of collateral damage which may result in creating new adversaries.
- *In a peacetime environment.* The commander must consider the political attitudes and previous actions of civilians in identifying hazards to friendly forces and the populace itself.

Adversaries are hostile elements other than the enemy that may be encountered during any operation. They present additional hazards. They may be organized opposition or individuals that challenge authority. They may include such diverse elements as rioters, criminals, rogues, or gangs that might want to harass a peace enforcement mission.

STEP 2. ASSESS HAZARDS

Step 2 completes the risk assessment. Risk is the chance of hazard or bad consequences. This step examines each hazard in terms of probability and severity to determine the risk level of one or more hazardous incidents that can result from exposure to the hazard. This step is conducted during three steps of the military decision-making process—*mission analysis, COA development,* and *COA analysis.* This step is also conducted after controls are developed.

The incident must be credible in that it must have a reasonable expectation of happening. The end result is an estimate of risk from each hazard and an estimate of the overall risk to the mission caused by hazards that cannot be eliminated. Leaders must also assess the risk to civilians posed by the operation. They may need to assess the operations' impact on the environment. This step is conducted in three substeps.

Substep A

Leaders and staffs assess each hazard in relation to the *probability* of a hazardous incident. The probability levels estimated for each hazard may be based on the mission, COAs being developed and analyzed, or frequency of a similar event. Figure 2-2 provides a summary of the five degrees of probability. The letters in parentheses following each degree (A through E) provide a symbol for depicting probability. For example, the letter *A* represents *frequent* probability.

FREQUENT (A) Occurs very often, continuously experienced	
Single item	Occurs very often in service life. Expected to occur several times over duration of a specific mission or operation. Always occurs.
Fleet or inventory of items	Occurs continuously during a specific mission or operation, or over a service life.
Individual soldier	Occurs very often in career. Expected to occur several times during mission or operation. Always occurs.
All soldiers exposed	Occurs continuously during a specific mission or operation.
LIKELY (B) Occurs several times	
Single item	Occurs several times in service life. Expected to occur during a specific mission or operation.
Fleet or inventory of items	Occurs at a high rate, but experienced intermittently (regular intervals, generally often,).
Individual soldier	Occurs several times in career. Expected to occur during a specific mission or operation.
All soldiers exposed	Occurs at a high rate, but experienced intermittently.
OCCASIONAL (C) Occurs sporadically	
Single item	Occurs some time in service life. May occur about as often as not during a specific mission or operation.
Fleet or inventory of items	Occurs several times in service life.
Individual soldier	Occurs some time in career. May occur during a specific mission or operation, but not often.
All soldiers exposed	Occurs sporadically (irregularly, sparsely, or sometimes).

Figure 2-2. Hazard Probability

SELDOM (D) Remotely possible; could occur at some time	
Single item	Occurs in service life, but only remotely possible. Not expected to occur during a specific mission or operation.
Fleet or inventory of items	Occurs as isolated incidents. Possible to occur some time in service life, but rarely. Usually does not occur.
Individual soldier	Occurs as isolated incident during a career. Remotely possible, but not expected to occur during a specific mission or operation.
All soldiers exposed	Occurs rarely within exposed population as isolated incidents.
UNLIKELY (E) Can assume will not occur, but not impossible	
Single item	Occurrence not impossible, but can assume will almost never occur in service life. Can assume will not occur during a specific mission or operation.
Fleet or inventory of items	Occurs very rarely (almost never or improbable). Incidents may occur over service life.
Individual soldier	Occurrence not impossible, but may assume will not occur in career or during a specific mission or operation.
All soldiers exposed	Occurs very rarely, but not impossible.

Figure 2-2. Hazard Probability (continued)

Substep B

Substep B addresses the *severity* of each hazard. It is expressed in terms of—

- Degree of injury or illness.
- Loss of or damage to equipment or property.
- Environmental damage.
- Other mission-impairing factors such as lost combat power.

The degree of severity estimated for each hazard may be based on knowledge of the results of similar past events. Figure 2-3 provides a summary of the four degrees of hazard severity. The Roman numerals in parentheses following each degree (I through IV) provide a convenient symbol for depicting severity. For example, *I* represents the *catastrophic* degree of severity.

CATASTROPHIC (I)	Loss of ability to accomplish the mission or mission failure. Death or permanent total disability (accident risk). Loss of major or mission-critical system or equipment. Major property (facility) damage. Severe environmental damage. Mission-critical security failure. Unacceptable collateral damage.
CRITICAL (II)	Significantly (severely) degraded mission capability or unit readiness. Permanent partial disability, temporary total disability exceeding 3 months time (accident risk). Extensive (major) damage to equipment or systems. Significant damage to property or the environment. Security failure. Significant collateral damage.
MARGINAL (III)	Degraded mission capability or unit readiness. Minor damage to equipment or systems, property, or the environment. Lost day due to injury or illness not exceeding 3 months (accident risk). Minor damage to property or the environment.
NEGLIGIBLE (IV)	Little or no adverse impact on mission capability. First aid or minor medical treatment (accident risk). Slight equipment or system damage, but fully functional and serviceable. Little or no property or environmental damage.

Figure 2-3. Hazard Severity

Substep C

In this substep leaders and staffs expand what they understand about probable hazardous incidents into estimates of levels of risk for each identified hazard and an estimate of the overall risk for the operation. Estimating risk follows from examining the outcomes of Substeps A and B; that is, both the probability and severity of hazardous incidents. This substep is more art than science. Much depends on the use of historical lessons learned, intuitive analysis, experience, and judgment. Uncertainty can arise in the assessment of both the probability and severity of a hazardous incident. Uncertainty results from unknowns about a situation; from incomplete, inaccurate, undependable, or contradictory information; and from unforeseen circumstances. Therefore, assessment of risk requires good judgment.

Figure 2-4 is a standardized matrix that can be used to assist in this process. Leaders and staffs enter the estimated degree of severity and probability for each hazard in Substeps A and B from the severity row and probability column, respectively. The point where the severity row and probability column intersect defines the level of risk. For example, if the hazard is estimated to have a *critical* severity (II) and a *likely* probability (B), the level of risk is high (H).

Figure 2-5 provides a summary of the levels of risk. It also provides examples of hazardous incidents for each risk level. Several examples illustrate the trade-off between tactical and accident risks.

Risk Assessment Matrix						
		Probability				
Severity		Frequent A	Likely B	Occasional C	Seldom D	Unlikely E
Catastrophic	I	E	E	H	H	M
Critical	II	E	H	H	M	L
Marginal	III	H	M	M	L	L
Negligible	IV	M	L	L	L	L

E – Extremely High Risk
H – High Risk
M – Moderate Risk
L – Low Risk

Figure 2-4. Risk Assessment Matrix

E - Extremely High: Loss of ability to accomplish the mission if hazards occur during mission. A *frequent* or likely probability of catastrophic loss (IA or IB) or *frequent* probability of *critical* loss (IIA) exists.

Example: A commander finds that one of his implied tasks to attack an objective involves crossing a normally shallow riverbed. After looking at the factors of METT-T, he discovers that three days of intense rain have raised the water level to rise above flood stage, with currents far in excess of his ability to safely ford with armored vehicles. After discussing COAs with his staff, he determines the accident risk is extremely high because of the likely probability and catastrophic severity of losing vehicles and killing soldiers. His conclusions are based on his experience with and knowledge of fording armored vehicles under the existing conditions of water depth and current speed.

H - High: Significant degradation of mission capabilities in terms of the required mission standard, inability to accomplish all parts of the mission, or inability to complete the mission to standard if hazards occur during the mission. *Occasional* to *seldom* probability of catastrophic loss (IC or ID) exists. A *likely* to *occasional* probability exists of a critical loss (IIB or IIC) occurring. *Frequent* probability of *marginal* losses (IIIA) exists.

Example: During a preplanned ambush, the leader discovers that the force he intends to ambush has significantly more combat power than his own force can accommodate. He realizes that he could only delay rather than destroy the enemy. He knows his casualty estimates would be very high if the enemy reorganized and counterattacked. He also knows that the size of the enemy force could seriously impact adjacent units conducting a movement to contact. He determines the situation is *high risk* because he estimates (based on his training and experience) there is a likely probability of the enemy reorganizing and counterattacking and the severity of loss to his unit would be critical.

M - Moderate: Expected degraded mission capabilities in terms of the required mission standard will have a reduced mission capability if hazards occur during mission. An *unlikely* probability of catastrophic loss (IE) exists. The probability of a *critical* loss is *seldom* (IID). *Marginal* losses occur with a *likely* or *occasional* probability (IIIB or IIIC). A *frequent* probability of negligible (IVA) losses exists.

Example: A commander in a defensive position receives a warning order to be prepared to counterattack if the enemy attacks again. He chooses to use pre-positioned ammunition caches to support his defense, as opposed to moving his ammunition resupply forward by truck. He determines that the severity of not having an immediate resupply of ammunition available during the counterattack will have a *critical* impact on his combat power. He realizes that if the enemy forces him to abandon his forward positions, the severity of the loss of his

Figure 2-5. Levels of Risk

ammunition caches will critically impact his combat power. He considers that his unit is deployed in excellent defensive positions. He has repelled two attacks that resulted in the destruction of an estimated 50 percent of the enemy's combat power. He receives information that the probability of the enemy attacking is *likely*, but that the probability of the enemy being reinforced and attacking in overwhelming force is remote *(seldom)*. The commander concludes that the risk of conducting a counterattack with limited ammunition is greater than the *moderate* risk of the enemy pushing him back.

L - Low: Expected losses have little or no impact on accomplishing the mission. The probability of *critical* loss is *unlikely* (IIE), while that of *marginal* loss is *seldom* (IIID) or *unlikely* (IIIE). The probability of a *negligible* loss is *likely* or *less* (IVB through (IVE).

Example: A mechanized task force (TF) conducting a movement to contact in a desert environment is overtaken by nightfall before reaching its limit of advance (LOA). The terrain along the axis of advance is flat and open. Visibility is about 800 meters under a clear sky illuminated by a full moon. Estimates put the enemy, which has been hastily withdrawing for the past three days, at approximately 30 percent strength. Contact has been light with no defensible terrain along the TF's axis. The TF commander considers all the factors. In addition, the TF is 100 percent operational in using night vision devices. The TF commander estimates that it is *unlikely* that his unit will incur losses of *critical* severity by being surprised by the enemy or lose *critical* combat power due to an accident. He estimates the risk to his force in continuing a nighttime movement is *low*.

Figure 2-5. Levels of Risk (continued)

STEP 3. DEVELOP CONTROLS AND MAKE RISK DECISIONS

Risk management is the recognition that decision making occurs under conditions of uncertainty. Decisions must remain consistent with the commander's stated intent and offer a good expectation of success. The risk-taking skill requires competency as a prerequisite.

FM 100-7, Decisive Force:
The Army in Theater Operations, May 1995

Step 3 is accomplished in two substeps: develop controls and make risk decisions. This is done during the COA development, COA analysis, COA comparison, and COA approval of the military decision-making process.

Substep A - Develop Controls

After assessing each hazard, leaders develop one or more controls that either eliminate the hazard or reduce the risk (probability and/or severity) of a hazardous incident. When developing controls, they consider the reason for the hazard not just the hazard itself.

Types of Controls. Controls can take many forms, but fall into three basic categories—*educational controls, physical controls,* and *avoidance.*

- *Educational controls.* These controls are based on the knowledge and skills of the units and individuals. Effective control is implemented through individual and collective training that ensures performance to standard.
- *Physical controls.* These controls may take the form of barriers and guards or signs to warn individuals and units that a hazard exists. Additionally, special controller or oversight personnel responsible for locating specific hazards fall into this category.
- *Avoidance.* These controls are applied when leaders take positive action to prevent contact with an identified hazard.

Criteria for Controls. To be effective, each control developed must meet the following criteria:

- *Suitability.* It must remove the hazard or mitigate (reduce) the residual risk to an acceptable level.
- *Feasibility.* The unit must have the capability to implement the control.
- *Acceptability.* The benefit gained by implementing the control must justify the cost in resources and time. The assessment of acceptability is largely subjective. Figure 2-6 gives criteria for determining acceptability of controls for each identified hazard.

Support	Availability of adequate personnel, equipment, supplies, and facilities necessary to implement a suitable controls.
Standards	Guidance and procedures for implementing a control are clear, practical, and specific.
Training	Knowledge and skills are adequate to implement a control.
Leadership	Leaders are competent to implement a control.
Individual	Individual soldiers are sufficiently self-disciplined to implement a control.

Figure 2-6. Criteria for Determining Acceptability of Controls

Examples of ControlsExamples of controls include—

- Engineering or designing to eliminate or control hazards.
- Selecting a COA that avoids identified hazards.
- Limiting the number of people and the amount of time they are exposed to hazards, consistent with mission requirements.
- Selecting personnel with appropriate mental, emotional, and physical capabilities.
- Providing protective clothing, equipment, and safety and security devices.
- Providing such services as adequate sanitation facilities and water purification capabilities.
- Providing warning signs and signals.
- Scheduling vehicle and aircraft silhouette drills.
- Planning training, including rehearsals, rock drills, battle drills, and so forth.
- Programming communications links for key civilian organizations.
- Establishing battlefield controls such as areas of operations and boundaries, direct fire control measures, fire support coordination measures, rules of engagement, airspace control measures, bridge classification, traffic control, and so forth.
- Developing terrorist attack warning systems and response plans.

The key is to specify who, what, where, when, and how each control is to be used. For example—

- Planning and scheduling intensive threat and friendly vehicle identification refresher training for all antiarmor and air defense weapons crews before the mission reduces the probability of engaging a friendly vehicle or aircraft (fratricide).
- Programming installation of crashworthy passenger seats in the UH-60 Blackhawk, when mission circumstances do not indicate their removal, can reduce the severity of injuries in crashes.
- Requiring soldiers to wear flak vests and helmets during movement to contact, or when riding in vehicles in areas where enemy fire is likely, can reduce the probability and severity of a wound from small arms fire or fragments.

- Establishing strong continuity documents and planning overlap tours for key leaders facilitate smooth transitions during extended operations.

Residual RiskOnce the responsible leader develops and accepts controls, he determines the residual risk associated with each hazard and the overall residual risk for the mission.

- *Residual risk* is the risk remaining after controls have been selected for the hazard. Residual risk is valid (true) only if the controls for it are implemented. As controls for hazards are identified and selected, the hazards are reassessed as in Step 2 and the level of risk is then revised. This process is repeated until the level of residual risk is acceptable to the commander or leader or cannot be further reduced. See Figures A-3 through A-5.
- *Overall residual risk* of a mission must be determined when more than one hazard is identified. The residual risk for each of these hazards may have a different level, depending on the assessed probability and severity of the hazardous incident. Overall residual mission risk should be determined based on the incident having the greatest residual risk. Determining overall mission risk by averaging the risks of all hazards is not valid. If one hazard has high risk, the overall residual risk of the mission is high, no matter how many moderate or low risk hazards are present.

Substep B - Make Risk Decision

A key element of the risk decision is determining if the risk is justified. The commander must compare and balance the risk against mission expectations. He alone decides if controls are sufficient and acceptable and whether to accept the resulting residual risk. If he determines the risk level is too high, he directs the development of additional controls or alternate controls, or he modifies, changes, or rejects the COA.

Leaders can use the risk assessment matrix in Figure 2-4—in conjunction with their commanders' guidance—to communicate how much risk they are willing to delegate. For example, a commander may place constraints on his subordinates that restrict their freedom of action to accept risk in instances where the risk might imperil his intent, his higher commander's intent, or a critical capability of the unit.

STEP 4. IMPLEMENT CONTROLS

Leaders and staffs ensure that controls are integrated into SOPs, written and verbal orders, mission briefings, and staff estimates. The critical check for this step, with oversight, is to ensure that controls are converted into clear, simple execution orders understood at all levels. Implementing controls includes coordination and communication with—

- Appropriate superior, adjacent, and subordinate units and those executing the mission.
- Logistics Civil Augmentation Program (LOGCAP) organizations and civilian agencies that are part of the force.

The media, NGOs, and PVOs must be included in coordination when their presence impacts or is impacted by the force.

Leaders must explain how supervisors will implement controls. Examples of control implementation include—

- Conducting vehicle and aircraft silhouette drills.
- Conducting rehearsals, rock drills, battle drills, and so forth.
- Conducting intensive threat and friendly vehicle identification refresher training for all antiarmor and air defense weapons crews.
- Conducting orientation for replacement personnel.
- Installing and maintaining communications links for key civilian organizations.
- Operating in convoys of four vehicles minimum.
- Carrying weapons and wearing flak jackets and helmets when outside secure compounds.

STEP 5. SUPERVISE AND EVALUATE

Leaders must supervise the execution of their orders. The more untrained the troops, the more detailed this supervision must be.

Infantry in Battle, 1939

During mission preparation and execution, leaders must ensure that their subordinates understand how to execute risk controls. Leaders continuously assess risks during the conduct of operations, especially during long-term missions. Leaders maintain situational awareness. They guard against complacency to ensure that risk control standards are not relaxed or violated. To gain insight into areas needing improvement, leaders must continuously evaluate their units' effectiveness in managing mission risks.

Supervise

Leaders supervise mission rehearsal and execution to ensure standards and controls are enforced. Techniques may include spot-checks, inspections, situation reports and brief-backs, buddy checks, and close supervision. During the mission, leaders continuously monitor controls to ensure they remain effective. They modify them as necessary. Leaders and individuals anticipate, identify, and assess new hazards to implement controls. They continually assess variable hazards such as fatigue, equipment serviceability, and the environment. Leaders modify controls to keep risks at an acceptable level.

During sustained operations, leaders continue planning to ensure that controls emplaced at the beginning of the mission apply to changes in the operation's current situation and to hazardous conditions. Leaders must maintain an extraordinary degree of discipline. They must avoid complacency, which can result from boredom and overconfidence. Leaders must ensure that soldiers do not relax their vigilance due to performing repetitive tasks—despite changing roles and missions, unit turbulence and turnover, and declining skills. Leaders maintain a close overwatch on controls put in place to reduce risks over a prolonged period. For example, during stability operations, land mine hazards may not be solved in the near term, but may require continual attention. Other examples of long-term hazards that may be encountered include—

- Climatic extremes.
- NBC and hazardous waste contamination.
- Diseases native to a particular area of operation or indigenous population.
- Terrorist threats.

Evaluate

After a mission, leaders and individuals evaluate how well the risk management process was executed. They—

- Determine how to ensure that successes are continued to the next mission.

- Capture and disseminate lessons learned so that others may benefit from the experience.
- Consider the effectiveness of the risk assessment in identifying and accurately assessing the probability and severity of hazards that resulted in mission degradation.
- Determine whether the level of residual risk of each hazard and of the overall mission were accurately estimated.
- Evaluate the effectiveness of each control in reducing or removing risk, including whether controls were effectively communicated, implemented and enforced.

Leaders and individuals determine why some controls were ineffective and what should be done when the hazard is encountered again. A control may be altered; the way it is implemented or supervised may be changed to make it effective; or a completely different control may be more effective. Leaders must energize the system to fix systemic problems that hinder combat effectiveness.

Figure 2-7 shows that the risk management process continues throughout a mission as well as from mission to mission. It is integral to the military decision-making process. Its application requires good judgment and intuitive analysis borne of confidence, experience, and situational awareness.

TOOLS AND PITFALLS

The appendix provides examples of risk management tools to help leaders assess identified hazards, develop controls, and make risk decisions. The tools should be tailored to suit particular situations and missions. The examples in Figures A-3 through A-5 are tools to manage risk at the tactical level. The example in Figure A-6 is a tool to manage risk at the operational level. Units may develop additional tools suitable for their needs.

Units train to a standard. They operate and train regardless of the degree of real or perceived difficulty. Risk reduction begins with commanders identifying their METLs. Commanders use the risk management process to assess the degree of risk related to each METL their unit must perform. From this assessment, risk reducing standard operating procedures evolve.

Figure A-7 provides an example of risk management considerations integrated into a mission training plan (MTP) task.

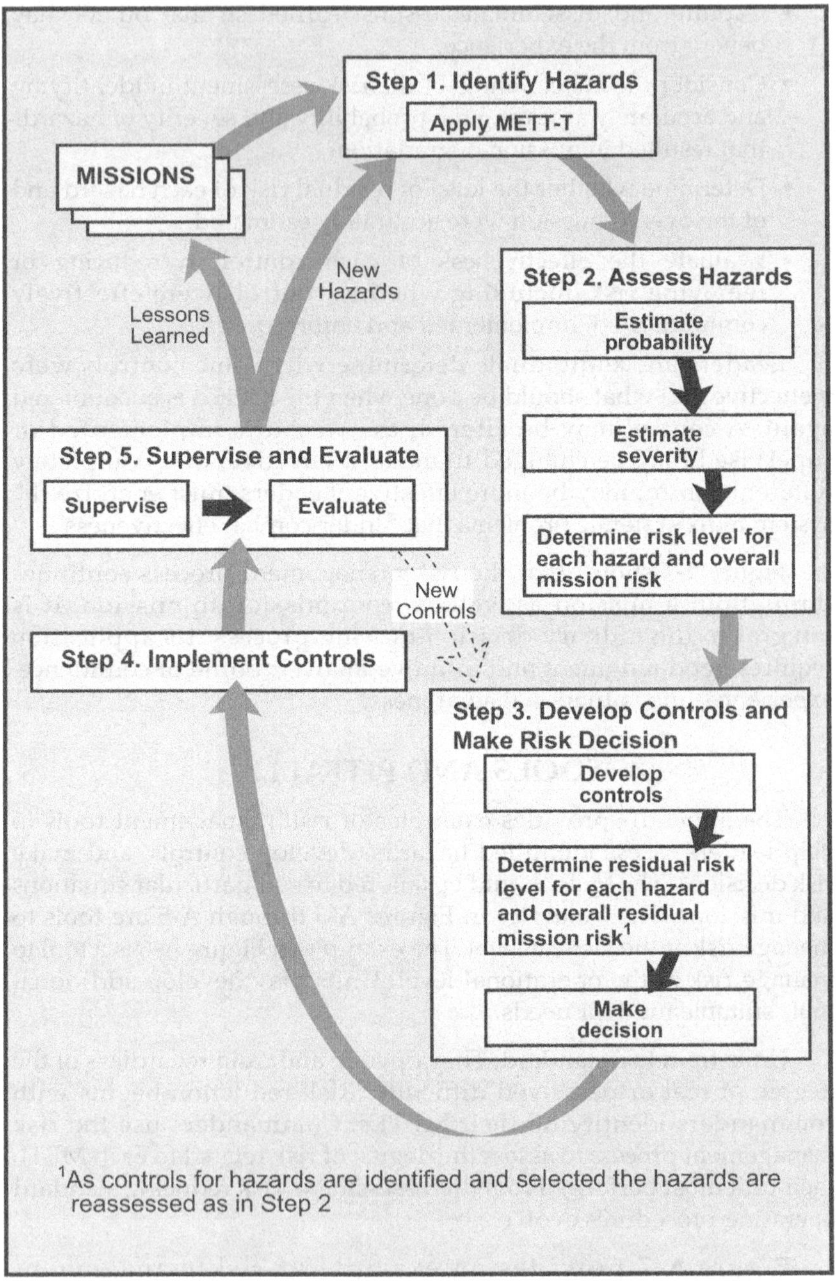

[1]As controls for hazards are identified and selected the hazards are reassessed as in Step 2

Figure 2-7. Continuous Application of Risk Management

Pitfalls arise when risk management tools are used without adaptation to the factors of METT-T. Using a standardized risk assessment card or checklist may be of some value initially in the mission analysis and COA development or in cases where a routine task is performed in an unchanging environment or static situation. However, such a tool used alone will not likely identify all hazards for every mission in a changing operational environment.

Completing the risk assessment alone, but failing to identify effective controls, usually results in a GO or NO-GO decision based on the initial risk. If the risk assessment does not accurately identify the hazards and determine the level of residual risk, the leader is likely to make his risk decision based upon incomplete or inaccurate information. If the risk assessment places missions in a routine, low-risk category, the commander may not be informed of a risk decision resulting in an accepted risk level that could imperil his or his higher commander's intent or other affected organizations. The risk management process is intended to provide reasonable controls to support mission accomplishment without exposing the force to unnecessary residual risk.

Chapter 3

Risk Management Implementation

It is imperative to develop twenty-first century leaders who trust their subordinates' abilities and judgment. Such leaders must be willing to underwrite their subordinates' honest errors and coach them on to excellence, without tolerating incompetence or laxity. We must recognize that Army leaders are not perfect, and that activity at the ragged edge of audacity sometimes leaves debris in its wake.

GEN William W. Hartzog
Commanding General
US Army Training and Doctrine Command

This chapter presents the moral and ethical implications of risk management. It outlines essential responsibilities and considerations necessary to effectively integrate and assess risk management within the Army. Leaders should tailor these essentials to develop specific how-to procedures suited to their circumstances, available resources, and mission.

MORAL AND ETHICAL IMPLICATIONS FOR LEADERS

To be successful, risk management must be underwritten by the chain of command. Leaders should not expect that all missions will be accomplished with zero defects—free from errors, flaws, or less-than-perfect performance. Demanding such rigid standards leads to oversupervision and paralysis; it produces timid leaders, afraid to make tough decisions in crisis and unwilling to take risks necessary for success in military operations. A zero defects mindset creates conditions that will lead inevitably, in the larger sense, to failure in battle and higher casualties. Leaders are morally bound to support a subordinate's decision to accept risks that are within his commander's intent and guidance, as he understands it.

Leaders accept that things may go wrong, even with the certain knowledge that a subordinate has done all within his power to prevent an incident. In such an event, the leader steps forward and accepts the responsibility along with his subordinate. Furthermore, risk management does not justify taking actions to facilitate an unethical or immoral action. FM 22-100 addresses the moral and ethical aspects of protecting the force within the overall framework of how to apply leadership at all levels to meet mission requirements.

> *Everyday as we respond to the nation's needs, we expose our soldiers to hazards in uncertain and complex environments. We do this with the full knowledge that there are inherent risks associated with any military operation. The nature of our profession will not allow for either complacency or a cavalier acceptance of risk.*
>
> General Dennis J. Reimer
> Chief of Staff, Army

RESPONSIBILITIES

One critical task for all operations is minimizing risk. Every military plan must make this a priority. It is an inherent part of every mission and a basic responsibility of commanders. Whether for training or operational deployments, commanders issue clear risk guidance. Minimizing risk—eliminating unnecessary risk—is the responsibility of everyone in the chain of command. This responsibility runs from the highest commander, through his subordinate leaders, to the soldier.

The commander and his staff must look at tactical risks and accident risks. They use the same risk management process to manage both types. Commanders—with the assistance of their leaders and staffs—manage accident risks. Commanders determine how and where they are willing to take tactical risks.

Leaders and soldiers at all levels are responsible and accountable for managing risks by ensuring that hazards and associated risks are—

- Identified during planning, preparation, and execution of operations.
- Controlled during preparation and execution of operations.

Soldiers are responsible for executing risk controls to standards. They must continuously assess variable hazards such as fatigue, equipment serviceability, and the environment. They must take care of one another and make risk decisions consistent with the higher commander's guidance.

Sometimes commanders are not properly advised in situations where the assumption of risk may affect or imperil their units, the intent of their higher commander, or the operations of an adjacent unit. This is most often attributed to—

- The *risk denial syndrome* in which leaders do not want to know of the risk.
- A staff member who believes that the risk decision is part of his job and does not want to bother his commander or section leader.
- A subordinate who does not fully understand the higher commander's guidance.
- Complacency—outright failure to recognize a hazard or the level of risk involved, or overconfidence in one's abilities or the unit's capabilities to avoid or recover from a hazardous incident.
- Use of a standardized risk assessment tool, such as a risk assessment card, that is not tailored to the unit's mission or adapted to the factors of METT-T and which may put missions in the routine low-risk category.

COMMANDERS

The commander directs the organization and sets priorities and the command climate (values, attitudes, and beliefs). Successful preservation of combat power requires embedding risk management into unit behavior. This requires commitment and creative leadership—innovative planning, careful management. It also requires the chain of command's demonstrated support of the risk management process. Only then will the Army begin to capture the full power of risk management. Commanders establish a command climate favorable for risk management integration by—

- Demonstrating consistent and sustained risk management behavior through leading by example—habitually doing risk management—and actively participating throughout the risk management process.
- Providing clear guidance, when appropriate, on where or what risk to accept.

- Obtaining and providing to subordinates the necessary assets to control risk.
- Knowing their own limitations, their leaders' and soldiers' limitations, and their unit's capabilities.
- Preventing a zero-defects mindset from creeping into their command's culture.
- Allowing subordinates to make mistakes and learn from them.
- Demonstrating full confidence in subordinates' mastery of their trade and their ability to execute a chosen COA.
- Keeping subordinates informed; consulting with subordinate leaders before making a decision, if feasible.
- Listening to subordinates.

Commanders are responsible and accountable for their own actions and those of units under their charge. Commanders must weigh the repercussions of casualties, damage to the environment, and loss of equipment. They must also consider the level of public reaction to loss against national, strategic, operational, or tactical objectives. Commanders are also responsible for keeping soldiers from falling into complacency. During sustained operations in particular, complacency can creep in as a result of accepting the status quo. For example, a unit can be deployed for several months and nothing may happen. The unit then gets very satisfied with itself and its performance. It gets conditioned to its initial circumstances—being well-armed, well-equipped, well-disciplined, and well-trained. It thinks it has risks under control and does not need to change. Such complacency, and the associated loss of situational awareness, can result in leaders and soldiers taking a gamble instead of a prudent risk. A risk is the accepted result of an informed decision; a gamble is an uninformed bet or guess on a hopeful outcome. Leaders and soldiers must clearly understand the difference.

Command is often exercised in conditions of uncertainty and ambiguity, where violence, danger, fear, and friction abound, and under the ever present time constraints driven by OPTEMPO. Risk decisions are frequently required by and dependent on the immediate situation. Judgment is required; a formula, rule, or checklist, by itself, is not appropriate under such circumstances.

Avoiding the zero-risk mindset requires the exercise of positive leadership. The commander's approach to managing risk should be through empowering leaders by pushing risk decisions as far down the chain of command as feasible within the next higher commander's guidance. Commanders must insist that subordinates exercise their freedom of action to act decisively and aggressively to complete assigned missions and promote success of the larger force. Subordinates must consider hazards outside their assigned responsibilities that impact the mission. The result is to encourage coordination and communication—laterally and up and down the chain of command. This requires and encourages initiative, which demands well-trained, determined, disciplined soldiers.

Risk management is a two-way street. It is important that those involved in mission preparation and execution be fully aware of the amount of command involvement and actions necessary to control or remove hazards. The higher commander's guidance specifies the degree of damage or risk to subordinate units that he is willing to accept during the current operation. Subordinates ensure they understand and implement their commander's intent and guidance. If, during the planning process, the accident and/or tactical residual risk exceeds that which the higher commander is willing to accept, the subordinate informs his commander. He requests the resources necessary to mitigate the risk. If, during mission execution, the subordinate determines the risk is too great, he directs the development of additional or alternate controls or modifies or changes the COA. He should notify his next higher commander of his decision. Requiring subordinates to report to the higher commander when a risk decision point is reached during mission execution can result in paralysis.

The objective of managing risk is not to remove all risk, but to eliminate unnecessary risk. Commanders conduct tough, realistic training, knowing that they may put lives and property at risk in the course of military operations. Nothing is worth the cost of a life as the result of taking unnecessary risk. If an action will result in an unacceptable risk, measures should be taken to mitigate it. If the risk cannot be mitigated to an acceptable level, the action should not be executed. Circumstances may occur during mission execution when a decision to stop and defer execution of the operation should be made to avoid taking unwarranted risk. Such a situation will generally occur at the tactical level. For example, circumstances may determine if a trade-off between maintaining the momentum of the attack or risking fratricide or serious accidents is justified. For example, during the deployment of TF Eagle to Bosnia-Herzegovina, there was pressure to complete the Sava River bridge as quickly as possible. The commander assessed the situation and decided to call for a short rest period because his soldiers were tired and getting careless. Consequently, they completed the bridge ahead of schedule under the most difficult conditions imaginable and without injuries.

LEADERS

Many of the tasks identified for commanders apply to all leaders. Leaders' specific responsibilities in managing risk include—

- Establishing clear, feasible risk management policies and goals.
- Conducting detailed planning within time constraints; assessing each mission and task in terms of its risk; continuously reassessing risk as the mission and conditions change and experience is gained.
- Making informed risk decisions and establishing and clearly communicating risk guidance.
- Training the risk management process. Ensuring subordinates understand the who, what, where, when, how, and why of managing risk and how the process applies to their circumstances and assigned responsibilities.
- Examining how subordinates manage risk and how soldiers protect themselves.
- Supervising and evaluating the unit's execution of risk controls during the mission to correct areas needing improvement.
- Advising his chain of command on risks and risk-reduction measures.
- Providing subordinates with feedback on their performance and ways to improve.
- Assessing the effectiveness of their unit's risk management program.
- Capturing and disseminating lessons learned to ensure they are continued from mission to mission so that others may benefit from the experience.

STAFFS

The chief of staff or executive officer is responsible for supervising integration of risk management across the staff. He coordinates development of risk reduction controls with emphasis on deconflicting controls that affect multiple functional areas and adjacent units. The staff officer helps the commander eliminate unnecessary risks by—

- Analyzing his functional area and applying risk management during the military decision-making process.
- Identifying constraints in the higher commander's risk guidance.
- Including hazards and their risks in the mission analysis briefing.
- Including a risk assessment for the commander's estimate.
- Considering the risk assessment in the operations estimate.
- Including risks and recommending ways to reduce their impact in the staff estimate.
- Implementing risk controls by coordinating and integrating them into the appropriate paragraphs and graphics of the OPORD and into products such as SOPs and OPLANs.
- Establishing procedures and standards that are clear and practical.
- Determining the effectiveness of hazard/risk controls and continuously assessing their suitability, feasibility, and acceptability.
- Supervising, evaluating, and assessing the integration of risk management during an operation.
- Continuously identifying hazards, assessing initial and residual risks for each hazard, recommending control measures to reduce the risk to the force.
- Identifying and assessing hazards associated with complacency, especially during extended operations, and recommending appropriate actions to the commander.

At the operational level, staffs focus on hazards and their risks across the spectrum of protecting the force

INDIVIDUALS

The level of expertise and maturity of the individual influence his proficiency in managing risk. Managing risk is subjective because its basis is individual judgment. Young soldiers are routinely charged with executing hazard controls and risk reduction measures. By nature, some them are impulsive risk takers. Their limited experience, coupled with an infallible attitude, can significantly increase the level of risk they are willing to accept. Their sense of indestructibility, motivation *(esprit de corps),* and willingness to achieve the mission at any cost also play a part.

Some soldiers and individuals ignore existing standards and improvise their own. Due to inexperience or complacency, they become susceptible to—

- Overestimating their ability to respond to or recover from a hazardous incident—they become overconfident.
- Underestimating the level of risk posed by a hazard.

It is imperative that individuals understand and execute controls implemented by leaders and staffs.

Individuals must maintain situational awareness and self-discipline when they perform their duties. They must—

- Understand and apply risk management.
- Execute controls directed by their leaders, that is, *perform to standards.*
- Carry risk management over into training and activities—both on and off duty.
- Look out for others—anyone has authority to halt something that is inherently unsafe.

INTEGRATION INTO TRAINING AND OPERATIONS

ARFOR commanders/leaders must continuously employ risk management approaches to effectively preclude unacceptable risks to personnel and property, including protecting forces preparing for or en route to combat.

FM 100-7, *Decisive Force: The Army in Theater Operations,* May 1995

Integrating risk management into training and operations—

- Preserves the lives and well-being of everyone.
- Conserves equipment, facilities, environmental resources, and combat power.

Risk management must not be treated as an afterthought. It must be planned for up front. Leaders and managers of materiel acquisition, base operations, and industrial operations must budget risk control costs up front at the level of expected payback over the duration of the activity, or the life cycle of materiel/weapons system.

When integrating risk management into sustained operations, leaders must consider increases in turbulence, personnel turnover, critical skill atrophy, and mission development. Leaders must continuously assess—

- The complexity of mission development and associated changing interrelationships with other agencies.
- The inclusion of civilian contractors, for example, LOGCAP, as part of the force.
- The presence of the media, NGOs, and PVOs.

These diverse elements need to be integrated into the risk management process.

Two key considerations relevant to managing risk in complex operational environments include—

- Understanding the culture of the indigenous population or society and its way of doing business. Leaders should respect their way of life and not interfere with local customs. Such interference could risk damage to relationships and increase the potential for introducing instability into the local society. Leaders must not, however, intentionally allow these considerations to endanger their force or its mission.
- The dynamics of managing risk and the way the leader and his subordinates manage risk. Leaders manage risk by—
  - — Having the right combination of well-trained, disciplined, well-armed, and well-equipped forces.
  - — Issuing clear guidance to minimize risk.
  - — Determining and implementing risk controls for carrying out the mission.

— Maintaining situational awareness, especially of vulnerabilities.

— Avoiding unnecessary accident risk.

Risk management works best when incorporated into existing command training and operational planning cycles. They then act as risk management tools and provide for—

- Development of policy, goals, objectives, and priorities in the commander's quarterly training guidance.
- The commander's training assessment, by identifying hazards and risk controls.
- Systematic observation and assessment of the unit's risk management performance and feedback into the training management cycle and SOPs.

Leaders and soldiers must have the skills, knowledge, and attitude to effectively manage risks inherent in all operations. Effective training helps soldiers become proficient. It qualifies them technically and tactically, and as leaders, to accomplish the mission without unnecessary risk.

Doctrine developers must integrate risk management into planning for all Army processes, especially the training management cycle defined in FMs 25-100 and 25-101. Unit leaders and their staffs must continually assess and evaluate the integration of risk management into short-, near-, and long-term training plans. They must continually review METLs to ensure that training is supported by realistic risk management objectives. In the past, unprepared or improperly trained units paid a high price for veteran status. The Army must learn from past experiences if it is to avoid repeating such losses.

ASSESSMENT OF THE RISK MANAGEMENT PROCESS

To assess the risk management process is to determine a unit's current level of proficiency in implementing the process. The term *assessment,* as discussed here, differs from *evaluation* as used in Step 5 of the process. Evaluation is used to measure demonstrated ability to accomplish specified objectives within a discrete operation or exercise. Assessment, as used here, also differs from the aspect of assessing hazards for probability and severity, as addressed in Step 2 of the process.

How well risk is managed affects readiness. Leaders need to know the current status and effectiveness of their organization's risk management program. They self-assess their unit's effectiveness in managing risk in order to gain insight into areas for improvement and get feedback on subordinates' understanding and application of risk guidance. The assessment objectives are to determine how—

- Effectively risk management is embedded into planning and preparing for operations.
- Well risk management is understood by subordinate leaders and soldiers.
- Effectively risk management is used to execute operations.

Leaders assess the effectiveness of their units by reviewing how well hazards are identified and risk controls are—

- Specified in oral and written OPORDs, OPLANs, and SOPs.
- Communicated to lowest level of chain of command.
- Included in short-, near-, and long-term training plans.
- Implemented into training and activities on and off duty.
- Embedded into protect-the-force programs such as safety and health and antiterrorism.
- Part of after-action reviews and fed into lessons learned.

Risk management cannot be seen as a competitive program whereby a unit or leader is judged or compared in a competitive sense. Focus is strictly on both reduction of risk and risk behavior.

Take calculated risks. That is quite different from being rash.

General George S. Patton, Jr.

Appendix

Examples of Risk Management Application

The examples in this appendix are designed to help those charged with managing risk.

TRACKING TOOL

The work sheet instructions are in Figure A-1.

Work Sheet Instructions	
Block	
A – D	Self explanatory
E	Identify task relating to the mission or task in Block A
F	**Identify Hazards** – Identify hazards by reviewing METT-T factors for the mission or task. Additional factors include historical lessons learned, experience, judgment, equipment characteristics and warnings, and environmental considerations.
G	**Assess Hazards** – Assessment includes historical lessons learned, intuitive analyses, experience, judgment, equipment characteristics and warnings, and environmental considerations. Determine initial risk for each hazard by applying risk assessment matrix (Figure 2-4). Enter the risk level for each hazard.
H	**Develop Controls** – Develop one or more controls for each hazard that will either eliminate the hazard or reduce the risk (probability and/or severity) of a hazardous incident. Specify who, what, where, why, when, and how for each control. Enter controls.
I	**Determine Residual Risk** – Determine the residual risk for each hazard by applying the risk assessment matrix (Figure 2-4). Enter the residual risk level for each hazard.
J	**Implement Controls** – Decide how each control will be put into effect or communicated to the personnel who will make it happen (written or verbal instruction; tactical, safety, garrison SOPs, rehearsals). Enter controls.
K	**Determine Overall Mission/Task Risk** – Select the highest residual risk level and circle it. This becomes the overall mission or task risk level. The commander decides whether the controls are sufficient to accept the level of residual risk. If the risk is too great to continue the mission or task, the commander directs development of additional controls or modifies, changes, or rejects the COA.
	Supervise and Evaluate – This last step is not on the worksheet. Plan how each control will be monitored for implementation (continuous supervision, spot-checks) and reassess hazards as the situation changes. Determine if the controls worked and if they can be improved. Pass on lessons learned.

Figure A-1. Risk Management Work Sheet Instructions

The work sheet (Figure A-2) provides a starting point to logically track the process of hazards and risks. It can be used to document risk management steps taken during planning, preparation, and execution of training and combat missions and tasks.

A. Mission or Task:	B. Date/Time Group Begin: End:	C. Date Prepared:
D. Prepared By: (Rank, Last Name, Duty Position)		

E. Task	F. Identify Hazards	G. Assess Hazards	H. Develop Controls	I. Determine Residual Risk	J. Implement Controls ("How To")

K. Determine overall mission/task risk level after controls are implemented (circle one)

LOW (L) MODERATE (M) HIGH (H) EXTREMELY HIGH (E)

SAMPLE

Figure A-2. Sample Risk Management Work Sheet

Examples provided in Figures A-3 through A-6 should help individuals manage risk at the tactical level.

A. Mission or Task: Prepare defensive positions			**B. Date/Time Group** **Begin:** 010035R May XX **End:** 010600R May XX		**C. Date Prepared:** 29 April XX
D. Prepared By: (Rank, Last Name, Duty Position) LT Jones, Plt Ldr					
E. Task	**F. Identify Hazards**	**G. Assess Hazards**	**H. Develop Controls**	**I. Determine Residual Risk**	**J. Implement Controls** ("How To")
Construct nonstandard antivehicular wire obstacle	Back injuries and wire cuts during materiel offload	Moderate (M)	Use proper lift and carry methods and wear concertina wire gloves and safety goggles	Low (L)	Unit TACSOP, ARTEP 5-145 DRILL (pg 2-44) ARTEP 5-335-11-MTP
	Blunt trauma and cuts in pounding of U-shaped pickets	Moderate (M)	Wear helmet and increase situational awareness	Low (L)	Unit TACSOP, ARTEP 5-145 DRILL (pg 2-44) ARTEP 5-335-11-MTP
	Cuts when unrolling concertina	Moderate (M)	Wear concertina wire gloves and maintain situational awareness	Low (L)	Unit TACSOP, ARTEP 5-145 DRILL (pg 2-44) ARTEP 5-335-11-MTP
	Cuts when installing concertina	Moderate (M)	Wear concertina wire gloves and maintain situational awareness	Low (L)	Unit TACSOP, ARTEP 5-145 DRILL (pg 2-44) ARTEP 5-335-11-MTP
	Cuts when installing barbed wire	Moderate (M)	Wear concertina wire gloves and maintain situational awareness	Low (L)	Unit TACSOP, ARTEP 5-145 DRILL (pg 2-44) ARTEP 5-335-11-MTP
K. Determine overall mission/task risk level after controls are implemented (circle one) LOW (L) (circled) MODERATE (M) HIGH (H) EXTREMELY HIGH (E)					

Figure A-3. Example of Completed Risk Management Work Sheet for Squad/Platoon

A. Mission or Task: Conduct a deliberate attack			B. Date/Time Group Begin: 010035R May XX End: 010600R May XX		C. Date Prepared: 29 April XX
D. Prepared By: (Rank, Last Name, Duty Position) CPT William Wallace, Cdr					
E. Task	**F. Identify Hazards**	**G. Assess Hazards**	**H. Develop Controls**	**I. Determine Residual Risk**	**J. Implement Controls** ("How To")
Conduct obstacle breaching operations	Obstacles	High (H)	Develop and use obstacle reduction plan	Low (L)	Unit TACSOP, OPORD, training handbook
	Inexperienced soldiers	High (H)	Additional instruction and increased supervision	Moderate (M)	Modified training schedule, additional instruction
	Operating under limited visibility	Moderate (M)	Use NVDs; use IR markers on vehicles	Low (L)	Unit TACSOP, OPORD
	Steep cliffs	High (H)	Rehearse use of climbing ropes	Moderate (M)	FM 90-6, *Mountain Operations* TC 90-6-1, *Mountaineering*
	Insufficient planning time	High (H)	Plan and prepare concurrently	Moderate (M)	OPORD, troop-leading procedures
K. Determine overall mission/task risk level after controls are implemented (circle one) LOW (L) MODERATE (M) [circled] HIGH (H) EXTREMELY HIGH (E)					

Figure A-4. Example of Completed Risk Management Work Sheet for Company/Team

A. Mission or Task: Redeploy unit to home station			B. Date/Time Group Begin: 010035R May XX End: 010600R May XX	C. Date Prepared: 29 April XX	
D. Prepared By: (Rank, Last Name, Duty Position) MAJ Woolsey, S3					
E. Task	**F. Identify Hazards**	**G. Assess Hazards**	**H. Develop Controls**	**I. Determine Residual Risk**	**J. Implement Controls** ("How To")
Conduct convoy operations-movement from Grafenwohr to home base	Fatigue leading to poor judgment/ accident-causing errors	Moderate (M)	1. Ensure all drivers receive adequate rest the night before movement 2. Brief planned rest stops and actions to take if driver becomes too tired/ill or has doubts about alertness 3. Have NCOs check physical alertness of drivers prior to movement	Low (L)	Unit SOP, OPORD on Tactical Road March, STP 21-2-MQS (Tactical Road March)
	Fast moving traffic mix of cars and large trucks with trailers	Moderate (M)	1. Use extreme caution when passing 2. Continue to scan (be aware of) traffic 3. Brief all drivers and vehicle commanders on lane-changing procedures	Low (L)	German laws and unit SOP
	Accidents and breakdowns blocking road	Moderate (M)	1. Brief drivers on actions to take a. Pull as far off the road as possible b. Get out on passenger side c. Post triangle sign at least 100 meters behind vehicle d. Stay as far away from vehicle and traffic as possible e. Brief all personnel on maintenance plan/action if vehicle is disabled f. Brief all personnel on procedures for vehicle exits on the traffic side	Low (L)	Unit SOP, OPORD on Tactical Road March

Figure A-5. Example of Completed Risk Management Work Sheet for Battalion/Task Force

A. Mission or Task: Redeploy unit to home station (continued)		B. Date/Time Group Begin: 010035R Dec XX End: 010600R Dec XX		C. Date Prepared: 29 Nov XX	
D. Prepared By: (Rank, Last Name, Duty Position) MAJ Woolsey, S3					
E. Task	**F. Identify Hazards**	**G. Assess Hazards**	**H. Develop Controls**	**I. Determine Residual Risk**	**J. Implement Controls ("How To")**
Conduct convoy operations-movement from Grafenwohr to home base	Accidents and breakdowns blocking road (continued)	Moderate (M)	2. Leaders ensure that proper maintenance are done prior to move (PMCS), dispatching and licensing procedure. 3. Contact team follows convoy and repairs or removes disabled vehicles from highway.	Low (L)	
	Weather conditions– high winds, sleet, snow	High (H)	1. Reduce speed according to environmental conditions less stopping distance available. 2. Road conditions especially right after rains – hydroplaning – review procedures for skids or hydroplanes. 3. Maintain convoy interval to allow for braking and assist civilian traffic in passing.	Moderate (M)	Unit SOP and vehicle operations manual
	High winds– reduction visibility, sudden reduction due to passing traffic	Moderate (M)	1. Change lanes carefully; ensure clearance and avoid erratic moves 2. Be aware of winds and effects on vehicles. 3. Drive defensively.	Low (L)	German traffic laws
	20% downgrade at checkpoint 1	Moderate (M)	1. Use lower gears. 2. Reduce speed to 25 mph	Low (L)	Unit SOP and vehicle operations manual

Figure A-5. Example of Completed Risk Management Work Sheet for Battalion/Task Force (continued)

A. Mission or Task: Redeploy unit to home station (continued)		B. Date/Time Group Begin: 010035R Dec XX End: 010600R Dec XX		C. Date Prepared: 29 Nov XX	
D. Prepared By: (Rank, Last Name, Duty Position) MAJ Woolsey, S3					
E. Task	**F. Identify Hazards**	**G. Assess Hazards**	**H. Develop Controls**	**I. Determine Residual Risk**	**J. Implement Controls** ("How To")
Conduct convoy operations-movement from Grafenwohr to home base	Heavy traffic, traffic jams, and congestion	Moderate (M)	1. Use extreme caution when passing 2. Use headlights at all times 3. Brief convoy speeds and vehicle intervals 4. Brief actions to take if vehicle pulls in front a. Maintain clearance from vehicle b. Maintain braking distance 5. Stay in right lane in construction zones 6. Ensure AMVs 2 or more meters wide remain in right lane 7. Continuously scan surroundings (situational awareness)	Low (L)	Provide risk assessment to convoy leaders. Provide safety brief information to leaders and drivers
K. Determine overall mission/task risk level after controls are implemented (circle one) LOW (L) MODERATE (M) [circled] HIGH (H) EXTREMELY HIGH (E)					

Figure A-5. Example of Completed Risk Management Work Sheet for Battalion/Task Force (continued)

A. Mission or Task: Peace Enforcement			**B. Date/Time Group** **Begin:** 010035R May XX **End:**		**C. Date Prepared:** 29 April XX
D. Prepared By: (Rank, Last Name, Duty Position) G-3 Div/Corps/EAC					
E. Task	**F. Identify Hazards**	**G. Assess Hazards**	**H. Develop Controls**	**I. Determine Residual Risk**	**J. Implement Controls** ("How To")
Occupy area of operations	Assault on facilities	Moderate (M)	Identify and isolate combatants, terrorists, supporters Harden potential terrorist targets	Low (L)	Coordination with local law enforcement agencies FM 5-103, *Survivability* *General /Order*
	Ambush	Moderate (M)	Kevlar helmets and flak jackets required outside cantonment area Minimum of 4 vehicles for movement Two qualified drivers per vehicle	Low (L)	Rehearsals Immediate action drills General orders
	Mines	High (H)	Countermine plan Mine awareness training Predeployment training	Moderate (M)	FM 20-32, *Mine/Countermine Operations* Lessons learned

Figure A-6. Example of Completed Risk Management Work Sheet for Division/Corps/EAC Task Force

A. Mission or Task: Peace Enforcement		B. Date/Time Group Begin: 010035R May XX End:		C. Date Prepared: 29 April XX	
D. Prepared By: (Rank, Last Name, Duty Position) G-3 Div/Corps/EAC					
E. Task	F. Identify Hazards	G. Assess Hazards	H. Develop Controls	I. Determine Residual Risk	J. Implement Controls ("How To")
Occupy area of operations	Fratricide	Moderate (M)	Identify friend/foe Antifratricide training LNO augmentations	Low (L)	Lessons learned ROE Weapons control procedures AOR orientation
	Seasonal risk (cold weather injuries)	Moderate (M)	Cold weather clothing and equipment Soldier/leader awareness training	Low (L)	Severe weather alert/action plan TB Med 507
	Seasonal risk (hot weather injuries)	Moderate (M)	Hot weather clothing and equipment Soldier/leader awareness training	Low (L)	Severe weather alert/action plan TB Med 507
	Disease	Moderate (M)	Immunizations Field sanitation Preventive medicine	Low (L)	Spring thaw plan Focused medical training Vector control program

Figure A-6. Example of Completed Risk Management Work Sheet for Division/Corps/EAC Task Force (continued)

A. Mission or Task: Peace Enforcement			B. Date/Time Group Begin: 010035R May XX End:		C. Date Prepared: 29 April XX
D. Prepared By: (Rank, Last Name, Duty Position) G-3 Div/Corps/EAC					
E. Task	**F. Identify Hazards**	**G. Assess Hazards**	**H. Develop Controls**	**I. Determine Residual Risk**	**J. Implement Controls** ("How To")
Occupy area of operations	Vehicle accidents	Moderate (M)	Driver qualifications Driver fatigue reduction Convoy safety procedures Monitor route conditions Vehicle inspections	Low (L)	Driver qualification training Driver sleep plans Two qualified drivers per vehicle Convoy safety briefings Route reconnaissance Vehicle inspections
K. Determine overall mission/task risk level after controls are implemented (circle one) LOW (L) MODERATE (M) [circled] HIGH (H) EXTREMELY HIGH (E)					

Figure A-6. Example of Completed Risk Management Work Sheet for Division/Corps/EAC Task Force (continued)

MISSION TRAINING PLAN

Figure A-7 shows risk management integrated into an MTP task. The example portrays a possible method for integrating the risk management process and products into MTPs. The text in bold italics indicates suggested areas in which to integrate risk management.

TASK: C^2 THE BATTALION (7-1-1901)(FM 7-20)

ITERATION: 1 2 3 4 5 (circle)

TRAINING STATUS: T P U (circle)

CONDITION: The brigade issues an OPORD or FRAGO

TASK STANDARD :

a. The battalion plan accomplishes the directed mission and specified tasks according to the brigade commander's concept and intent. The plan is received and understood by the leadership of the battalion, who makes the plan successful. It is coordinated with higher, adjacent, and supporting elements.

b. The plan is as fully prepared as time allows to initiate the mission at the directed time.

c. The battalion controls and synchronizes subordinate and supporting elements so that it accomplishes the mission and preserves the force. ***Include commander's risk guidance.***

d. The battalion keeps higher, adjacent, subordinate, supporting, and supported headquarters informed of essential information key to controlling the battle or making required decisions.

SUBTASKS AND STANDARDS: GO NO-GO

+1. Battalion leaders issue the warning order.

a. A complete warning order is issued within 15 minutes of receipt of the brigade order.

b. Warning order is received by all platoons within 45 minutes of issuance of battalion warning order.

+ Critical task.

* Leader task.

Figure A-7. Integration of Risk Management into MTP Task

*2. Battalion commander analyzes mission and gives initial guidance.

a. Guidance includes restated mission, which includes brigade commander's intent for battalion and identifies all specified and implied tasks.

b. Guidance is given within 30 minutes of receipt of order.

c. Guidance includes instructions on information requirements and initially required preparation actions (movement, resupply) to start. ***Guidance includes chain of command authorized to accept risk (extremely high, high, moderate, and low) affecting higher command's intent, resources, or adjacent units' missions.***

+3. Battalion accomplishes reconnaissance and other actions to gather needed information.

a. Reconnaissance actions begin to physically gain information on the enemy and terrain as early as possible.

b. Commander conducts a personal reconnaissance, when possible. If not, the commander conducts a detailed map reconnaissance.

c. Subordinate leaders perform a personal reconnaissance when possible. See subordinate company (ARTEP 7-10-MTP) and platoon (ARTEP 7-8-MTP) T&EOs.

d. Staff coordinates with subordinate, higher, supporting, and adjacent headquarters to gather information for planning. ***Staff analyzes mission to identify hazards.***

e. Staff provides operations, intelligence, and CSS estimates to include all critical METT-T factors and ***risk assessment considerations (identified hazards/risks).***

+4. Battalion commander develops and wargames courses of action and selects one.

Figure A-7. Integration of Risk Management into MTP Task (continued)

a. Tactically feasible courses of action (include CS and maneuver) are made and wargamed with the available staff (commander, S3, and FSO are best for quick planning sequences; XO, S2, engineer, S4, S3 (Safety), and ADA officer are best in more deliberate situations). ***Each course of action contains assessment of hazards, risk level, and control measures identified to lower or control the risk.***

b. Best COA is selected.

c. COA is wargamed and refined by the command and staff. The staff must understand the concept to produce a sound OPORD and rehearse. ***The staff continually assesses the situation to identify new hazards, assess the risk from each hazard, develop acceptable control measures, and reassess for residual risk to the mission. COA should pose minimum risk to soldiers, equipment, and mission accomplishment. Risk acceptance decisions for the COA are made at the appropriate level in the chain of command.***

*5. Staff develops an OPLAN and OPORD from the commander's guidance.

OPLAN and OPORD successfully accomplish the mission according to higher commander's intent. ***Hazards and risk control measures are included in the appropriate paragraphs and graphics.***

*6. Battalion commander and staff issue implementing FRAGOs (accident risk controls).

*7. Battalion commander issues the OPORD/FRAGO.

a. OPORD/FRAGO is issued IAW the one-third, two-thirds rule and makes full use of daylight time.

b. OPORD/FRAGO accomplishes all directed missions and tasks, complies with the brigade commander's intent, controls risk, and is doctrinally sound. It is based on evaluator judgment and on comparison of brigade and battalion OPORDs.

Figure A-7. Integration of Risk Management into MTP Task (continued)

c. All subordinate and supporting elements receive the OPORD/FRAGO.

d. OPORD/FRAGO contains task organization, mission, concept, ***accident risk controls,*** intent for maneuver, supporting fires, and obstacles; missions and tasks for each subordinate; fire support and CSS instructions; and coordinating instructions to synchronize the efforts of maneuver forces and CS.

e. If more time is available, the battalion commander issues a fully developed OPORD (although an initial FRAGO may be issued to allow subordinates to begin preparation and followed by a full OPORD.)

f. Order is given at a location that reduces travel time, allows observation of the zone/sector, and promotes OPSEC. (Depending upon the METT-T factors, observation of the zone/sector may not be possible.)

g. Battalion commander should perform brief-backs and war gaming, ***to include identifying hazards and control measures,*** immediately after the order to improve subordinate understanding and reaction.

h. Subordinate leaders and staff should perform lateral coordination before leaving the orders site.

+8. Commander and staff coordinate and refine the plan.

a. Time is well used to continue gathering information and to improve the plan (contingency plans, ***hazard identification and controls,*** fire plans) .

b. New information is disseminated and coordinated with higher, adjacent, and supporting headquarters to include—

- Changes or refinements in plan.
- Information on the enemy in the sector or zone.

Figure A-7. Integration of Risk Management into MTP Task (continued)

- Information that impacts on planning and execution ***(risk acceptance decisions/hazard controls)*** of subordinate elements.
- Adjustments/changes in the plans.

+*9. Battalion executes changes in task organization.

a. Main CP coordinates link-up location, time, and responsible element.

b. Attachments/new elements are received at the coordinated location and time; updated on current situation, OPORDs, and SOIs; and resupplied.

c. Detachments reach the link-up point at the time and place directed.

+*10. Battalion performs and commander and staff perform, supervise, and monitor preparations.

a. Command group/XO performs brief-backs with subordinate commanders, leaders, and key staff.

b. Main CP maintains status of preparations.

c. Elements make full use of time to prepare for the operation. Subjective judgment of the evaluator is based on the analysis of preparation charts and available preparation time.

+*11. Battalion sees the battlefield.

a. Command group is positioned to see and move.

b. Companies and other subordinates accurately report critical information on actions and changes in combat status within five minutes. See subordinate MTPs.

c. Main CP collects, analyzes, and passes processed critical information.

d. Subordinates execute intelligence collection plan. See

Figure A-7. Integration of Risk Management into MTP Task (continued)

subordinate element MTPs. ***Subordinates integrate the risk management process when developing plans and executing tasks.***

+*12. Battalion leaders command and control the execution.

a. Subordinate elements report enemy and friendly actions, change in status, and any other factor that would require change within three minutes.

b. Battalion leaders win the battle by directing the maneuver of units, controlling direct and indirect fires, ***properly integrating risk management into planning, preparation, and execution,*** and directing other CS actions to cope with new METT-T/risk factors. Indicators are:

(1) Elements not following OPLAN and OPORD are corrected.

(2) Responses to new METT-T hazards are directed soon after the new situation occurs.

(3) No friendly casualties inflicted by friendly direct or indirect fires or other accidents.

(4) Number/percentage of direct fire weapons engaging the enemy.

(5) Number/percentage of indirect fire rounds fired and percentage hitting/suppressing the enemy.

(6) Number of enemy casualties.

(7) Number of friendly casualties.

c. The C^2 and CSS assets are controlled to support maneuver effort. Indicators are—

(1) Effective CSS and C^2.

(2) C^2 or CSS elements not destroyed by enemy direct fires.

d. FRAGOs are clear, concise, ***include risk management,*** and are quickly executed by subordinates.

Figure A-7. Integration of Risk Management into MTP Task (continued)

e. Changes that affect the battle are disseminated within five minutes.

+13. Subordinate commanders, leaders, and staff laterally coordinate actions during the battle.

All battle actions requiring coordination between elements are coordinated.

+*14. Battalion coordinates with adjacent and supporting headquarters.

All battle actions requiring coordination with other headquarters are laterally and promptly coordinated.

+*15. Battalion reports.

Battalion CPs submit all critical and required reports to brigade. They report events to adjacent and supporting elements that impact on them in time for those units to react. ***They advise the chain of command, as appropriate, on risks and risk-reduction measures.***

Figure A-7. Integration of Risk Management into MTP Task (continued)

Glossary

AAR after-action review

ADA air defense artillery

ARFOR army forces headquarters

ARTEP Army Training and Evaluation Program

assessment an analytical process to determine an organization's current levels of proficiency on a specific objective (for example, a training objective or risk management implementation) (CJCSM 3500.03)

base operations support the provision of administrative and logistical services; includes supply operations, maintenance of materiel, personnel support, base services and administrative services rendered by or through activities of the supporting installation

BASOPS base operations support

C^2 command and control

COA course of action

CofS chief of staff

combat power the total means of destructive and/or disruptive force that a military unit or formation can apply against an opponent at a given time; a combination of the effects of maneuver, firepower, protection, and leadership

controls actions taken to eliminate hazards or reduce their risk

CP command post

CSS combat service support

CTC combat training center

danger exposure or vulnerability to harm or risk; the balance between the chance or probability of a hazardous incident and the result of the hazardous incident

EAC echelons above corps

evaluation the process used to measure the demonstrated ability to accomplish specified objectives such as training within a discrete event or exercise (CJCSM 3500.03); measurement of the demonstrated ability of soldiers or units to perform a task and supporting skill and knowledge or learning objective against the established standard

exposure the frequency and length of time personnel and equipment are subjected to a hazard

FM field manual

FRAGO fragmentary order

fratricide the employment of friendly weapons and munitions with the intent to kill the enemy or destroy his equipment or facilities, which results in unforeseen and unintentional death or injury to friendly personnel

friction the accumulation of chance errors, unexpected difficulties, enemy actions, and confusion of battle

FSO fire support officer

G3 general staff operations section

hazard any actual or potential condition that can cause injury, illness, or death of personnel, damage to or loss of equipment, property or mission degradation (FM 101-5); a condition or activity with potential to cause damage, loss or mission degradation (Joint Pub 1-02)

inherently dangerous an activity or task containing a danger to life or limb that is a permanent and inseparable element of the activity

IPB intelligence-preparation-of-the-battlefield

IR infrared

LOA limit of advance

LOGCAP Logistics Civil Augmentation Program

METL mission-essential task list

METT-T mission, enemy, terrain, troops, and time available

MOS military occupational specialty

MTP mission training plan

NBC nuclear, biological, chemical

NVD night vision device

OCOKA O - observation and fields of fire, C - cover and concealment, O - obstacles, K - key terrain and decisive terrain, A - avenues of approach

OPCON operational control

operational tempo the pace of an operation or operations; OPTEMPO includes all of the activities the unit is conducting; OPTEMPO can be a single activity or a series of operations

OPLAN operations plan

OPORD operations order

OPTEMPO operational tempo

personnel tempo unit work load level and number of deployed days per year

PERSTEMPO personnel tempo

PIR priority intelligence requirements

probability the likelihood that a hazardous incident will occur

PVO private voluntary organization

residual risk the level of risk remaining after controls have been identified and selected for hazards that may result in loss of combat power

risk chance of hazard or bad consequences; the probability of exposure to chance of injury or loss from a hazard; risk level is expressed in terms of hazard probability and severity (FM 101-5)

risk assessment identification and assessment of hazards (first two steps of risk management process); an identified hazard is assessed to determine the risk (both the probability of occurrence and resulting severity) of a hazardous incident due to the presence of the hazard

risk decision the decision to accept or not accept the risks associated with an action; made by the commander, leader, or individual responsible for performing that action

risk management the process of identifying, assessing, and controlling risks arising from operational factors and making decisions that balance risk cost with mission benefits

risk management integration the embedding of risk management principles and practices into Army operations, culture, organizations, systems, and individual behavior

S3 operations officer

severity the expected consequence of an event (hazardous incident) in terms of degree of injury, property damage, or other mission-impairing factors (loss of combat power and so on) that could occur

situational awareness ability to have accurate and real-time information on friendly, enemy, neutral, and noncombatant locations; a common, relevant picture of the battlefield scaled to specific level of interest and special need

SOP standing operating procedure

T&EO test and evaluation outline

TACSOP tactical standing operating procedure

TC training circular

TF task force

US United States

XO executive officer

References

SOURCES USED

FM 20-400. *Military Environmental Protection.* 1997 (draft).

FM 22-9. *Soldier Performance in Continuous Operations.* 12 December 1991.

FM 22-100. *Army Leadership.* 1997 (draft).

FM 25-100. *Training the Force.* 15 November 1988.

FM 25-101. *Battle-Focused Training.* 30 September 1990.

FM 34-60. *Counterintelligence.* 5 February 1990.

FM 34-130. *Intelligence Preparation of the Battlefield.* 23 May 1989.

FM 100-5. *Operations.* August 1997 (draft).

FM 100-40. *Tactics.* 29 October 1997 (draft).

FM 100-7. *Decisive Force: The Army in Theater Operations.* 31 May 1995.

FM 101-5. *Staff Organization and Operations.* 31 May 1997.

READINGS RECOMMENDED

Army Regulation 70-1. *Systems Acquisition Policy and Procedures.* 1997.

Army Regulation 385-16. *System Safety Engineering and Management.* 3 May 1990.

CSA Statement on Risk Management. 27 July 1995.

HQDA Letter 5-97-1. Risk Management Integration Responsibilities. 1 May 1997.

MIL-STD-882C. *System Safety Program Requirements.* 19 January 1993.

"Risk Management for Brigades and Battalions." *Center for Army Lessons Learned (CALL) Newsletter 95-9.* June 1995.

Made in the USA
Monee, IL
04 November 2026

67382844R00046